SpringerBriefs in Molecular Science

Ultrasound and Sonochemistry

Series Editors

Bruno G. Pollet, Faculty of Engineering, Norwegian University
of Science and Technology, Trondheim, Norway
Muthupandian Ashokkumar, School of Chemistry, University
of Melbourne, Parkville, VIC, Australia

SpringerBriefs in Molecular Science: Ultrasound and Sonochemistry
is a series of concise briefs that present those interested in this broad
and multidisciplinary field with the most recent advances in a broad
array of topics. Each volume compiles information that has thus far
been scattered in many different sources into a single, concise title,
making each edition a useful reference for industry professionals,
researchers, and graduate students, especially those starting in a new
topic of research.

More information about this subseries at http://www.springer.com/
series/15634

About the Series Editors

Bruno G. Pollet is a full Professor of Renewable Energy at the Norwegian University of Science and Technology (NTNU) in Trondheim. He is a Fellow of the *Royal Society of Chemistry* (RSC), an Executive Editor of *Ultrasonics Sonochemistry* and a Board of Directors' member of the *International Association of Hydrogen Energy* (IAHE). He held Visiting Professorships at the University of Ulster, Professor Molkov's HySAFER (UK) and at the University of Yamanashi, Professor Watanabe's labs (Japan). His research covers a wide range of areas in Electrochemical Engineering, Electrochemical Energy Conversion and Sono-electrochemistry (Power Ultrasound in Electrochemistry) from the development of novel materials, hydrogen and fuel cell to water treatment/disinfection demonstrators & prototypes. He was a full Professor of Energy Materials and Systems at the University of the Western Cape (South Africa) and R&D Director of the National Hydrogen South Africa (HySA) Systems Competence Centre. He was also a Research Fellow and Lecturer in Chemical Engineering at The University of Birmingham (UK) as well as a co-founder and an Associate Director of The University of Birmingham Centre for Hydrogen and Fuel Cell Research. He has worked for Johnson Matthey Fuel Cells Ltd (UK) and other various industries worldwide as Technical Account Manager, Project Manager, Research Manager, R&D Director, Head of R&D and Chief Technology Officer. He was awarded a Diploma in Chemistry and Material Sciences from the Université Joseph Fourier (Grenoble, France), a B.Sc. (Hons) in Applied Chemistry from Coventry University (UK) and an M.Sc. in Analytical Chemistry from The University of Aberdeen (UK). He also gained his Ph.D. in Physical Chemistry in the field of Electrochemistry and Sonochemistry under the supervision of Profs. J. Phil Lorimer & Tim J. Mason at the Sonochemistry Centre of Excellence, Coventry University (UK). He undertook his PostDoc in Electrocatalysis at the Liverpool University Electrochemistry group led by Prof. David J. Schiffrin. Bruno has published many scientific publications, articles, book chapters and books in the field of Sonoelectrochemistry, Fuel Cells, Electrocatalysis and Electrochemical Engineering. Bruno is member of editorial board journals (*International Journal of Hydrogen Energy/Electrocatalysis/Ultrasonics Sonochemistry/Renewables-Wind, Water and Solar/Electrochem*). He is also fluent in English, French and Spanish. *Current Editorships: Hydrogen Energy and Fuel Cells Primers Series* (*AP, Elsevier*) and *Ultrasound and Sonochemistry* (*Springer*).

Prof. Muthupandian Ashokkumar (Ashok) is a Physical Chemist who specializes in Sonochemistry, teaches undergraduate and postgraduate Chemistry and is a senior academic staff member of the School of Chemistry, University of Melbourne. Ashok is a renowned sonochemist, with more than 20 years of experience in this field, and has developed a number of novel techniques to characterize acoustic cavitation bubbles and has made major contributions of applied sonochemistry to the Materials, Food and Dairy industry. His research team has developed a novel ultrasonic processing technology for improving the functional properties of dairy ingredients. Recent research also involves the ultrasonic synthesis of functional nano- and biomaterials that can be used in energy production, environmental remediation and diagnostic and therapeutic medicine. He is the Deputy Director of an Australian Research Council Funded Industry Transformation Research Hub (ITRH; http://foodvaluechain.unimelb.edu.au/#research; Industry Partner: Mondelez International) and leading the Encapsulation project (http://foodvaluechain.unimelb.edu.au/research/ultrasonic-encapsulation). He has received about $ 15 million research grants to support his research work that includes several industry projects. He is the Editor-in-Chief of *Ultrasonics Sonochemistry*, an international journal devoted to sonochemistry research with a Journal Impact Factor of 4.3). He has edited/co-edited several books and special issues for journals; published ~ 360 refereed papers (H-Index: 49) in high impact international journals and books; and delivered over 150 invited/keynote/plenary lectures at international conferences and academic institutions. Ashok has successfully organised 10 national/international scientific conferences/workshops and managed a number of national and international competitive research grants. He has served on a number of University of Melbourne management committees and scientific advisory boards of external scientific organizations. Ashok is the recipient of several prizes, awards and fellowships, including the Grimwade Prize in Industrial Chemistry. He is a Fellow of the RACI since 2007.

Bruno G. Pollet · Muthupandian Ashokkumar

Introduction to Ultrasound, Sonochemistry and Sonoelectrochemistry

 Springer

Bruno G. Pollet
Faculty of Engineering
Norwegian University of Science
and Technology
Trondheim, Sor-Trondelag, Norway

Muthupandian Ashokkumar
School of Chemistry
University of Melbourne
Parkville, VIC, Australia

ISSN 2191-5407   ISSN 2191-5415 (electronic)
SpringerBriefs in Molecular Science
ISSN 2511-123X   ISSN 2511-1248 (electronic)
Ultrasound and Sonochemistry
ISBN 978-3-030-25861-0   ISBN 978-3-030-25862-7 (eBook)
https://doi.org/10.1007/978-3-030-25862-7

This Springer imprint is published by the registered company Springer Nature Switzerland AG
The registered company address is: Gewerbestrasse 11, 6330 Cham, Switzerland

Contents

About the Authors

Bruno G. Pollet is a full professor of renewable energy at the Norwegian University of Science and Technology (NTNU) in Trondheim and an extraordinary professor at the University of the Western Cape (UWC) in Cape Town. He is Fellow of the *Royal Society of Chemistry* (RSC, UK), Executive Editor of *Ultrasonics Sonochemistry* and Board of Directors member of the *International Association of Hydrogen Energy* (IAHE, USA). His research covers a wide range of areas in electrochemical engineering, electrochemical energy conversion and sonoelectrochemistry (power ultrasound in electrochemistry) from the development of novel materials, hydrogen and fuel cell to water treatment/disinfection demonstrators and prototypes. He was awarded a Baccalauréat série C, Grenoble, France, a Diploma in Chemistry and Material Sciences from the Université Joseph Fourier, Grenoble, France, a B.Sc. (Hons) in applied chemistry from Coventry University, England, and an M.Sc. in analytical chemistry from The University of Aberdeen, Scotland. He also gained his PhD in physical chemistry in the field of electrochemistry and sonochemistry under the supervision of profs. J. Phil Lorimer and T. J. Mason at the Sonochemistry Centre of Excellence, Coventry University, England. He has published more than 150 papers and edited a number of books—*Current Editorships: Hydrogen Energy and Fuel Cells Primers Series* (*AP, Elsevier*) and *Ultrasound and Sonochemistry* (*Springer*).

Muthupandian Ashokkumar is a full professor of chemistry at The University of Melbourne, Australia. Within the school, he is the group leader of the sonochemistry research group. He obtained his Bachelor of Science (1982) and Master's (1984) degrees from Madurai Kamaraj University, Madurai, India. In 1989, he obtained his PhD from the University of Madras, Chennai, India. His research group has been exploring various industrial applications of ultrasound, primarily the use of ultrasound in food processing, materials synthesis and biomedical applications. The sonochemistry research group has also been investigating the use of ultrasound for synthesising targeted drug delivery agents. In addition, they use ultrasound to generate efficient catalytic materials for wastewater treatment using alternative energy resources, such as solar energy. Ashokkumar is the recipient of a number of awards, including the Grimwade Prize in Industrial Chemistry. He is editor-in-chief of *Ultrasonics Sonochemistry* journal, has published more than 380 papers and edited a number of books.

Chapter 1
Fundamental and Applied Aspects of Ultrasonics and Sonochemistry

1.1 Introduction

Ultrasound refers to sound waves of frequency greater than 20 kHz. Based on the nature of applications, it is generally classified into two ranges: high-frequency low-intensity (HFLI) ultrasound and low-frequency high-intensity (LFHI) ultrasound. HFLI ultrasound ranges from 1 to 10 MHz, and its power is generally in the range of milliwatts. Within this range, sonication does not produce much physical or chemical effects, therefore is usually used in medical diagnostic applications. For this reason, it is also known by the name *diagnostic ultrasound* [1]. LFHI ultrasound is in the range 20 kHz–1 MHz and can be operated at thousands of watts, therefore also known as *power ultrasound* [2]. This category of ultrasound is used for many applications such as cleaning, emulsification, processing, extraction and chemical reactions [3–10].

Generation of ultrasound requires the use of a transducer, a device that is able to convert electrical energy into sound energy [11]. The most common transducer material is lead zirconate titanate (PZT), a ceramic ferroelectric crystal with strong piezoelectricity. When alternating electric field is supplied to the transducer, the PZT crystal undergoes expansion and compression. This resonance oscillation

© The Author(s), under exclusive licence to Springer Nature Switzerland AG 2019
B. G. Pollet and M. Ashokkumar, *Introduction to Ultrasound, Sonochemistry and Sonoelectrochemistry*, Ultrasound and Sonochemistry
https://doi.org/10.1007/978-3-030-25862-7_1

frequency is specific to the dimension of the PZT crystal; therefore, each transducer is built for a specific frequency. When this oscillating crystal is in contact with a liquid system, the oscillation is transmitted to the system in the form of sine waves (ultrasound).

Sonication of a liquid medium generates chemical and physical effects [2–10]. The chemical effects of ultrasound do not come directly from the interaction of the acoustic waves with the molecules; instead, the effects come from acoustic cavitation process. By definition, cavitation refers to creating a cavity in a liquid medium. However, the cavitation process involves the formation, growth and the collapse of a microbubble in an ultrasonically irradiated solution.

1.2   Acoustic Cavitation

The medium through which ultrasound travels through experiences alternating acoustic pressure cycles. During rarefaction, the negative pressure causes a pulling effect causing bubble nucleation. In this case, the cohesive force between solvent molecules needs to be overcome to create a bubble/cavity. For cavitation bubble to form, a high negative pressure is required. Theoretical negative pressure required to form a cavity in water is of the order of a few hundred atmospheres [12]. However, experiments have shown that the actual pressure required is much less than the theoretical value. This is because of the advantage of having impurities, dissolved gases and particulates in water. Once a bubble/cavity is formed, it undergoes oscillations under the pressure variation as discussed below.

The Rayleigh–Plesset equation [13] is one of the mostly well-accepted theoretical models for describing bubble radial dynamics as well as bubble collapse. Parameters such as fluid compressibility, damping, condensation and evaporation are included in the theory. One of the derivation processes applied was the work done by an oscillating bubble (W_b) to the surrounding fluid medium as shown in Eq. (1.1):

$$W_b - W_l = E_k \tag{1.1}$$

where E_k is the kinetic energy of liquid volume around the bubble and W_l is the work done by the liquid volume to the surrounding liquid if considered as incompressible during bubble expansion. These three elements in Eq. (1.1) can be expressed individually as follows:

$$E_k = \frac{1}{2}\rho \int_{R_i}^{R_l} \dot{r}^2 4\pi r^2 dr = 2\pi\rho R_i^3 \dot{R}^2 \tag{1.2}$$

$$W_b = \int_{R_0}^{R_i} 4\pi r^2 p_l dr \tag{1.3}$$

$$W_l = \int_{R_0}^{R_i} 4\pi r^2 p_\infty dr \tag{1.4}$$

where ρ is the density of the liquid, R_l is the radius of the liquid volume, R_i is the instantaneous bubble radius $(R_l \gg R_i)$, r is the radial distance from the bubble centre (dot stands for the time derivative), R_0 is the ambient bubble radius; p_l is the liquid pressure, and p_∞ is the pressure on the surface of the liquid including the acoustic pressure. Equation (1.5) can be derived by substitution of Eqs. (1.2)–(1.4) in Eq. (1.1) with respect to R while applying the condition shown in Eq. (1.6).

$$\frac{p_l - p_\infty}{\rho} = \frac{3\dot{R}^2}{2} + R\ddot{R} \tag{1.5}$$

$$\frac{\partial(\dot{R}^2)}{\partial R} = \frac{1}{\dot{R}}\frac{\partial(\dot{R}^2)}{\partial t} = 2\ddot{R} \tag{1.6}$$

where $\dot{R}$ and $\ddot{R}$ are the velocity and acceleration of the bubble, respectively. Several key parameters such as bubble-solution interfacial tension, Poisson's constant (heat capacity ratio) of the dissolved

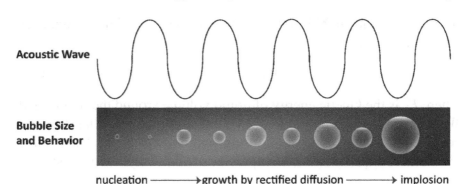

Fig. 1.1 Net growth of a microbubble by rectified diffusion from the act of ultrasound waves

gas in liquid medium, viscosity of the liquid medium and the frequency of ultrasound can be derived from Eq. (1.5) on the left side. Detailed studies on bubble dynamics are available in a number of review articles [14–17].

Bubble oscillations result in the growth of bubble size under specific experimental conditions. The growth of the microbubbles occurs through rectified diffusion [18–20] as shown in Fig. 1.1.

Rectified diffusion is "the slow growth of a pulsating gas bubble due to an average flow of mass into the bubble as a function of time". The rectification of a microbubble can be explained by two effects: area and shell effects [18]. According to the area effect, the surface area and size of a microbubble undergo changes due to the expansion and compression cycles. During the expansion cycle, the gas and vapour molecules from bulk solution evaporate and diffuse into the bubble due to the very low internal pressure. During the compression cycle, the gas and vapour molecules are being expelled out due to the high internal pressure. The amount of gas/vapour molecules that diffuse in during expansion phase is relatively larger due to higher surface area of the bubble wall in its expanded phase resulting in a net growth of the bubble as shown in Fig. 1.1. On the other hand, the shell effect pays more attention to the gas concentration within a liquid shell around the bubble. The diffusion rate is dependent on the concentration gradient between the bubble wall and the edge of the liquid shell. A combination of area and shell effects results in the rectified growth of bubbles over many acoustic cycles. The growth continues until it reaches its resonance size, a point that the bubble oscillation frequency is in the same range as the driving ultrasound frequency. The relationship is given by Eq. (1.7), where f is the acoustic frequency in hertz and R is the bubble's radius in metres [21].

$$f \times R \approx 3 \tag{1.7}$$

At its resonance frequency, the microbubble experiences a large pulsation resulting in the implosion of the bubble, which is a near adiabatic process. The physical conditions after the collapse of the cavitation bubble can be estimated using Eqs. (1.8) and (1.9) [21].

$$T_{max} \approx p_g \left(\frac{p_{\infty(\gamma-1)}}{p_g}\right)^{\frac{\gamma}{\gamma-1}} \tag{1.8}$$

$$P_{max} \approx T_0 \left(\frac{p_{\infty(\gamma-1)}}{p_g}\right) \tag{1.9}$$

where p_g is the pressure of the gas inside the cavitation bubble. The estimated localised temperature and pressure can reach up to 5000 K and 150 MPa depending upon various conditions of the oscillating bubble.

1.3   Bubble Temperature

It should be noted that Eq. (1.8) overestimates T_{max} since it does not take into account the heat leaking from the bubble or thermal conductivity of the gases or the energy consumed in endothermic chemical reactions within the bubble. More elaborate models for bubble collapse exist, and they do consider the "missing" aspects of Eq. (1.8) [21]. Experimental determinations of the temperature within a cavitation bubble have been made by a number of research groups. By fitting the experimentally recorded single bubble sonoluminescence spectra using the spectrum radiated by a blackbody, temperatures in the range of 5000 to >50,000 K [21] have been estimated. However, experimental estimates of the temperature within the collapsing bubbles based on multibubble sonochemistry and sonoluminescence are reported to be between 750 and 6000 K [22–30].

Sonolysis of methane in argon saturated water has been used by Henglein and co-workers [24] to estimate the bubble core temperature. The approach taken by Henglein is the dependence of rate constants on temperature (Arrhenius equation). Methane, being volatile, evaporates into the cavitation bubble and pyrolysed under high-temperature conditions generated within the bubble. One of the pyrolysis products is methyl radical that undergo recombination reactions by two pathways as shown in Fig. 1.2. The rate constant for ethane formation is less temperature-dependent, whereas the rate constant for ethylene formation is highly temperature-dependent (see Eqs. (1.10) and (1.11) and insert in Fig. 1.2).

$$CH_3 \cdot + CH_3 \cdot \rightarrow C_2H_6$$
$$k = 2.4 \times 10^{14} * T^{-0.4} \text{ cm}^3 \text{ mol}^{-1} \text{ s}^{-1} \tag{1.10}$$

$$CH_3 \cdot + CH_3 \cdot \rightarrow C_2H_4 + H_2$$
$$k = 1.0 \times 10^{16} * \exp(-134{,}000 \text{ J/RT}) \text{ cm}^3 \text{ mol}^{-1} \text{ s}^{-1} \tag{1.11}$$

The relationship between the ratio between the rate constants and temperature is shown in Fig. 1.2. The ratio between rate constants is directly related to the yields of ethane and ethylene (plus acetylene). Hence, it is possible to estimate the bubble temperature by experimentally measuring the yields of ethane, ethylene and acetylene.

Depending upon the percentage of methane and argon dissolved in water, the temperature was estimated to be in the range of 1930–2720 K (the temperature decreased with an increase in the percentage of the methane). Tauber et al. [25] used a similar technique in their study of the sonolysis of t-butanol in water and estimated the

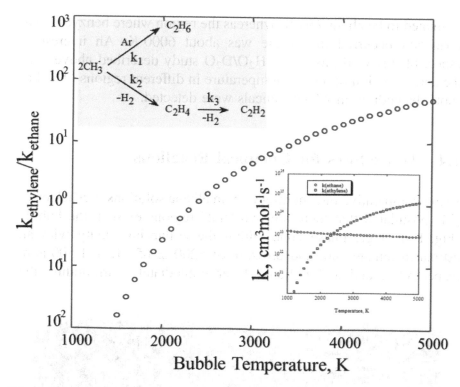

Fig. 1.2 Methyl radical recombination method to estimate bubble temperature

temperature to be in the range 2300–3600 K. The lower temperature was obtained for solution with the highest concentration of t-butanol (0.5 M). Both these studies show that the core temperature of the bubble is strongly dependent upon the amount of organic solute in the system.

Misík and Riesz [26], using the kinetic isotope effect in an EPR spin-trapping study of the sonolysis of H_2O/D_2O mixtures, found that the cavitation temperature determined was dependent on the specific spin trap used and fell in the range of 1000–4600 K. They attributed the difference in the values obtained by the spin traps as being due to the sampling of different regions of the cavitation "hot spot". Misík and Riesz [26] made use of the kinetic isotope effect in the ultrasound-induced production of radicals in organic liquids to esti-mate the temperatures during cavitation. The temperature region where hydrogen radical abstraction occurred in *n*-dodecane was estimated to be about 750 K, whereas the region where benzyl radical formation occurred in toluene was about 6000 K. An interesting aspect of this work, as in the H_2O/D_2O study described above, was the determination of a mean temperature in different regions of a "hot spot" depending on which radicals were detected.

1.4   Active Sites for Chemical Reactions

Using comparative rate thermometry in alkane solutions, Suslick et al. [31] postulated that there are two "hot" regions exist in the bubble (Fig. 1.3): a gas-phase zone within the collapsing cavity with an estimated temperature and pressure of 5200 ± 650 K and 500 atm, respectively, and a thin liquid layer immediately surrounding the

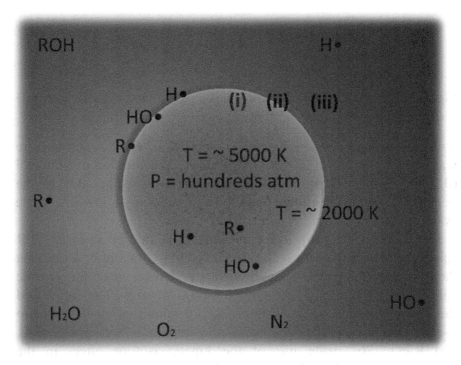

Fig. 1.3 Sonochemical reaction sites. (i) Inner cavitation bubble, (ii) shell of the microbubble and (iii) bulk solution

collapsing cavity with an estimated temperature of ~ 1900 K. It is also of relevance to mention that characteristic emission bands have been observed from a number of transition metal atoms (SL) formed within the bubbles, by the sonolysis of volatile metal carbonyls in organic liquids [31, 32]. The relative intensities and line broadening of the emission bands observed have also been used to calculate the temperature (5000 K) and pressure (500 atm) generated by cavitation.

Another important aspect that is schematically shown in Fig. 1.3 is that chemical reactions occur in three distinct regions during acoustic cavitation. The first region is the hot interior of the bubble where maximum temperature is attained on bubble collapse that generates highly reactive radicals from volatiles solutes and gas molecules that diffused into the bubble during the expansion phase. For example, in aqueous solutions, water molecules decompose to hydrogen and hydroxide radicals [Reaction (R.1)]. They are known as the primary radicals. The radicals can react among each other to form hydrogen gas, hydrogen peroxide or back to water molecules [Reactions (R.2)–(R.4)]. Other reactions occur based on the gas content of the bubble as shown in Reactions (R.5)–(R.8). If surface active molecules are present, secondary radicals are formed when the primary radicals attack them. For an example, when alcohol is present, different alcohol or alkyl radicals may form [Reactions (R.9) and (R.10)]. There have been many studies that reported on the importance of both the primary and secondary radicals in chemical reactions [33–40].

$$H_2O \rightarrow H + OH \tag{R.1}$$

$$2OH \rightarrow H_2O_2 \tag{R.2}$$

$$2H \rightarrow H_2 \tag{R.3}$$

$$H + OH \rightarrow H_2O \tag{R.4}$$

$$O_2 \rightarrow 2O \tag{R.5}$$

$$O + O_2 \rightarrow O_3 \tag{R.6}$$

$$\mathbf{H} + O_2 \rightarrow \mathbf{HO_2} \tag{R.7}$$

$$2\mathbf{HO_2} \rightarrow H_2O_2 + O_2 \tag{R.8}$$

$$H + CH_3CH_2OH \rightarrow CH_3CHOH + H_2 \tag{R.9}$$

$$OH + CH_3CH_2OH \rightarrow CH_3CHOH + H_2O \tag{R.10}$$

The second region (Fig. 1.3 (ii)) is reported to have temperature around 2000 K during bubble collapse. While the importance of interfacial chemistry in sonochemical reactions has been extensively reported in the literature [3, 41], recent studies by Cavalieri and co-workers have demonstrated the advantages of the hot interfacial region of cavitation bubbles for synthesising polyphenolic nanostructures [42–44]. They have provided experimental evidence that oligomeric phenolic structures are produced due to the combined effect of "hot interface" and oxidative radicals generated during cavitation using N-benzoyl-L-tyrosine ethyl ester (BTTE) as a model compound. As shown in Fig. 1.4, oxidation of the phenolic moiety of BTTE by OH radicals followed by the formation of biphenyl links leads to the formation of BTTE oligomers, which then self-assemble to generate nanoparticles possessing biofunctionality.

The third region (Fig. 1.3 (iii)) is the bulk solution of a sonicated system. Solutes which are not surface active remain in bulk solution [3, 34]. In such cases, the redox radicals generated within cavitation bubbles diffuse into the bulk and react with such solutes. A typical example is the formation of metal nanoparticles by the reaction between metal ions in solution and H atoms generated within the bubbles [Reactions (R.11) and (R.12)] due to pyrolysis of water molecules [Reaction (R.1)]. Figure 1.5 shows the formation of gold colloids when an aqueous solution of gold chloride ($AuCl_4^-$) is sonicated.

$$AuCl_4^- + 3H\cdot \rightarrow Au^0 + 4Cl^- + 3H^+ \tag{R.11}$$

$$nAu^0 \rightarrow Au_n \tag{R.12}$$

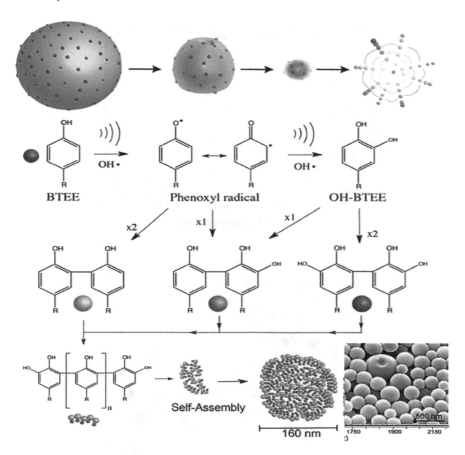

Fig. 1.4 Combined effect of hot bubble–solution interface and oxidative radicals generated during acoustic cavitation to generate biofunctional nanoparticles. Adapted from Cavalieri et al. [42]

1.5 Physical Effects

In addition to the effect of high-temperature conditions in promoting chemical reactions, the high pressures generated within cavitation bubbles play an equally important role. For example, the high pressure is released into the medium as shock waves. It is strong enough to disturb the molecular structure of the substrate present and change

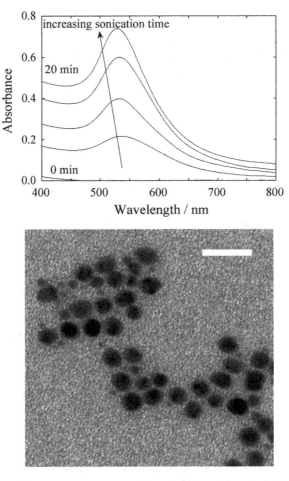

Fig. 1.5 Above: Absorption spectrum showing the formation of gold nanoparticles (plasmon and centred around 530 nm). Below: Electron microscopic images of gold nanoparticles; scale bar = 50 nm)

their reactivity [3–5]. The chemical reactivity in a system can also be enhanced by the shear forces generated due to bubble oscillations, bubble collapse and shock waves. Shock waves are also known to play major role in emulsification. It could break the liquid–liquid boundary in a heterogeneous system, resulting in the formation of a homogeneous system. As mentioned before, shock waves can also cause erosion. Another interesting phenomenon is the asymmetrical collapse of the microbubble resulting in the formation of microjets. It can also accelerate chemical activity and causes erosion, fragmentation and pitting. In a sonicated system, apart from the shear caused by bubble implosion, shear is also generated during bubble oscillation by acoustic microstreaming. With the size and volume change during pulsation, the surrounding liquid is being pulled and pushed away repeatedly, resulting in turbulence. This shear also helps in the mechanical mixing and emulsification processes. The rapid movement helps in enhancing chemical reactivity.

1.6 Sonoluminescence and Sonochemiluminescence

The implosion of cavitation bubbles under specific experimental conditions results in the emission of light, which is known as sonoluminescence [45, 46]. This phenomenon is directly correlated to the temperature produced during cavitation. A number of theories exist to describe sonoluminescence [46]. If luminol is present in the solution, HO· radicals may react with luminol and generate an intermediate that can chemiluminesce. This is referred to as sonochemiluminescence [47]. When ultrasound waves travel from the transducer through the medium, it can be reflected once it reaches a reflector, such as the container's wall or the air–water boundary. When this happens, the reflected waves travel in the opposite direction resulting in the formation of a standing wave containing stationary nodes and antinodes. Such phenomenon can be seen in from sonochemiluminescence imaging technique Fig. 1.6.

In Fig. 1.6a, the blue-coloured regions represent antinodes, where active cavitation bubbles are located. The formation of active microbubbles at the antinodes is also schematically shown in

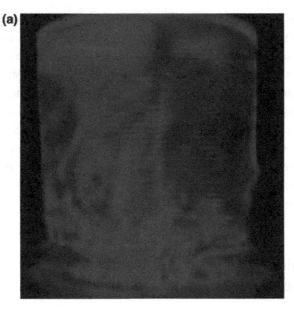

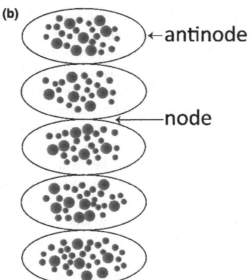

Fig. 1.6 a Antinodes and nodes of a sonicated luminol solution are visible in the sonochemiluminescence (SCL) image. **b** The chemiluminescing microbubbles are observed at the antinodes

Fig. 1.6b. This method is very useful in gaining visual understanding of a cavitation profile in a sonicated system.

1.7 Applied Sonochemistry

The generation of chemical and physical effects during acoustic cavitation has been discussed in the previous section. Both effects have been utilised to carry out numerous chemical reactions and chemical processing applications [3–8]. For example, oxidising radicals/molecular species (**OH** and HO_2 and H_2O_2) have been used for the degradation of organic pollutants in aqueous environment [39]. Most of such environmental applications of sonochemistry are laboratory-scale studies. However, some pilot studies have been reported on the environmental applications of sonochemistry [3]. The primary (**H**) and secondary (alkyl, alcohol) reducing radicals generated during acoustic cavitation have been used for synthesising metal, inorganic and bio-nanomaterials for various applications [3, 31–34]. Again, most of these studies are laboratory-scale proof of concept studies.

The physical effects such as shock waves and microjets have been used in several processing applications as well as increasing the rate of chemical reactions by enhanced mass transfer effects. In the past decade or so, the use of physical effects of ultrasound/sonication for food processing applications has received significant attention from researchers and food industry. For example, the functionality of dairy proteins is significantly influenced by the mass transfer effects generated during sonication. Ultrasonic extraction and emulsification have been developed for large-scale processing applications [3–8].

Unique applications of ultrasonics and sonochemistry where both the physical and chemical effects are utilised include emulsion polymerisation and biomaterials synthesis. Since these topics have been covered in separate publications of this series (*Springer Briefs in*

Molecular Science, Ultrasound and Sonochemistry), readers are encouraged to refer to these books for further reading.

In summary, this chapter provides a simple overview of the basic aspects of sonochemistry, namely how acoustic cavitation process is generated and how the physical and chemical effects generated by acoustic cavitation have been used in a variety of applications. The following chapter focuses on the electrochemical applications of sonochemistry.

References

1. Thrush A, Martin K, Hoskins PR (eds) (2010) Diagnostic ultrasound. Cambridge University Press, UK
2. Gallego-Juarez JA, Graff KF (eds) (2017) Power ultrasonics: applications of high-intensity ultrasound. Elsevier Science and Technology, UK
3. Ashokkumar M, Cavalieri F, Chemat F, Okitsu K, Sambandam A, Yasui K, Zisu B (eds) (2016) Handbook of ultrasonics and sonochemistry, vol 1 and 2. Springer Reference, Singapore
4. Chen D, Sharma SK, Mudhoo A (eds) (2012) Handbook on applications of ultrasound: sonochemistry for sustainability. CRC Press, USA
5. Feng H, Weiss J, Barbosa-Cánovas G (eds) (2011) Ultrasound technologies for food and bioprocessing. Springer, New York
6. Povey MJW, Mason TJ (1998) Ultrasound in food processing. Blackie Academic, London
7. Feng H, Barbosa-Cánovas GV, Weiss J (2011) Ultrasound technologies for food and bioprocessing. Springer, New York
8. Mason TJ (ed) (1999) Advances in sonochemistry, vol 5. Elsevier
9. Adam MI, Dobiás P, Eisner A, Ventura K (2009) Extraction of antioxidants from plants using ultrasonic methods and their antioxidant capacity. J Sep Sci 32:288–294
10. Mason TJ (1988) Sonochemistry. Oxford Chemistry Premiers
11. Safari A, Akdogan EK (eds) (2008) Piezoelectric and acoustic materials for transducer applications. Springer
12. Ashokkumar M, Mason T (2007) Sonochemistry, Kirk-Othmer encyclopedia of chemical technology. Wiley, USA

13. Rayleigh L (1917) On the pressure developed in a liquid during the collapse of a spherical cavity. Lond Edinb Dubl Phil Mag 34:94–98
14. An Y (2012) Nonlinear bubble dynamics of cavitation. Phys Rev E 85:016305
15. Yasui K (1999) Single-bubble and multibubble sonoluminescence. Phys Rev Lett 83:4297
16. Plesset M (1949) The dynamics of cavitation bubbles. J Appl Mech 16:277–282
17. Brenner MP, Hilgenfeldt S, Lohse D (2002) Single bubble sonoluminescence. Rev Mod Phys 74:425
18. Crum L (1984) Acoustic cavitation series: part five rectified diffusion. Ultrasonics 22:215–223
19. Lee J, Kentish SE, Ashokkumar M (2005) Effect of surfactants on the rate of growth of an air bubble by rectified diffusion. J Phys Chem B 109:14595–14598
20. Leong T, Wu S, Kentish S, Ashokkumar M (2010) Growth of bubbles by rectified diffusion in aqueous surfactant solutions. J Phys Chem C 114:20141–20145
21. Leighton T (1994) The acoustic bubble. Academic Press, London
22. Flannigan DJ, Suslick KS (2012) Temperature nonequilibration during single-bubble sonoluminescence. J Phys Chem Lett 3:2401–2404
23. Pflieger R, Ouerhani T, Belmonte T, Niketenko SI (2017) Use of NH (A3Π–X3Σ−) sonoluminescence for diagnostics of nonequilibrium plasma produced by multibubble cavitation. Phys Chem Chem Phys 19:26272–26279
24. Hart EJ, Fischer C-H, Henglein A (1990) Sonolysis of hydrocarbons in aqueous solution. Int J Radiat Appl Instrum C, Radiat Phys Chem 36:511–516
25. Tauber A, Mark G, Schuchmann H-P, von Sonntag C (1999) Sonolysis of tert-butyl alcohol in aqueous solution. J Chem Soc Perkin Trans 2:1129–1136
26. Mišík V, Riesz P (1996) EPR study of free radicals induced by ultrasound in organic liquids II. Probing the temperatures of cavitation regions. Ultrason Sonochem 3:25–37
27. Ciawi E, Rae J, Ashokkumar M, Grieser F (2006) Determination of temperatures within acoustically generated bubbles in aqueous solutions at different ultrasound frequencies. J Phys Chem B 110:13656–13660
28. Rae J, Ashokkumar M, Eulaerts O, von Sonntag C, Reisse J, Grieser F (2005) Estimation of ultrasound induced cavitation bubble temperatures in aqueous solutions. Ultrason Sonochem 12:325–329

29. Ciawi E, Ashokkumar M, Grieser F (2006) On the limitations of the methyl radical recombination method for acoustic bubble temperature measurements in aqueous solutions. J Phys Chem B 110:9779–9781
30. Ashokkumar M, Grieser F (2005) A comparison between multibubble sonoluminescence intensity and the temperature within cavitation bubbles. J Am Chem Soc 127:5326–5327
31. Suslick KS, Hammerton DA, Cline RE (1986) Sonochemical hot spot. J Am Chem Soc 108:5641–5642
32. Suslick KS, Price GJ (1999) Applications of ultrasound to materials chemistry. Ann Rev Mater Sci 29:295–326
33. Ohayon E, Gedanken A (2010) The application of ultrasound radiation of the synthesis of nanocrystalline metal oxide in a non-aqueous solvent. Ultrason Sonochem 17:173–178
34. Ashokkumar M (2008) Sonochemical synthesis of inorganic nanoparticles. In: Cozzoli PD (ed) Advanced wet-chemical synthetic approaches to inorganic nanostructures, Chap. 4. Transworld Research Network, pp 107–131
35. Mason TJ, Peters D (2002) Practical sonochemistry, power ultrasound uses and applications, 2nd edn. Ellis Horwood Publishers, Chichester
36. Neppolian B, Doronila A, Grieser F, Ashokkumar M (2009) Simple and efficient sonochemical method for the oxidation of arsenic(III) to arsenic (V). Environ Sci Technol 43:6793–6798
37. Okitsu K, Sharyo K, Nishimura R (2009) One-pot synthesis of gold nanorods by ultrasonic irradiation: the effect of pH on the shape of the gold nanorods and nanoparticles. Langmuir 25:7786–7779
38. Babu SG, Ashokkumar M, Neppolian B (2016) The role of ultrasound on advanced oxidation processes. In: Sonochemistry: from basic principles to innovative applications. Topics Curr Chem 374(Article number 75)
39. Petrier C, Francony A (1997) Ultrasonic waste-water treatment: incidence of ultrasonic frequency on the rate of phenol and carbon tetrachloride degradation. Ultrason Sonochem 4:295–300
40. Gedanken A (2008) Preparation and properties of proteinaceous microspheres made sonochemically. Chem A Eur J 14:3840–3853
41. Ashokkumar M, Grieser F (2007) The effect of surface active solutes on bubbles in an acoustic field. PhysChemChemPhys 9:5631–5643
42. Cavalieri F, Colombo E, Nicolai E, Rosato N, Ashokkumar M (2016) Sono-assembly of nanostructures via tyrosine–tyrosine coupling reactions at the interface of acoustic cavitation bubbles. Mat. Horizons 3:563–567
43. Bhangu SK, Ashokkumar M, Cavalieri F (2017) A simple one-step ultrasonic route to synthesize antioxidant molecules and fluorescent nanoparticles from phenol and phenol-like molecules. ACS Sustain Chem Eng 5:6081–6089

44. Zhu H, Cavalieri F, Ashokkumar M (2018) Ultrasound-assisted synthesis of cross-linked poly(ethylene glycol) nanostructures with hydrophobic core and hydrophilic shell. Macromol Chem Phys 219(1–5):1800353
45. Young FR (1999) Cavitation. World Scientific
46. Young FR (2005) Sonoluminescence. CRC Press, NY
47. Hatanaka SI, Mitome H, Yasui K, Hayashi S (2002) Single-bubble sonochemi-luminescence in aqueous luminol solutions. J Am Chem Soc 124:10250–10251

Chapter 2
Short Introduction to Sonoelectrochemistry

2.1 Introduction

By definition, sonication or more accurately ultrasonication is the application of vibrational energy in the ultrasonic frequency range of (20 kHz–1 MHz) to efficiently mix aqueous solutions and suspensions [1]. Historically, low-frequency high-intensity ultrasound has been commonly used in the food and medical sectors for the homogenisation and emulsification of biological materials such as proteins and lipids, and sterilisation of foodstuffs. It is well known that ultrasonic waves generated by the ultrasonic transducer (ultrasonicator, in the form of an ultrasonic probe, also called a *cell disruptor*, or an ultrasonic bath—see Fig. 2.1) are able to disrupt the lipid bilayers of cell membranes [2]. For chemical applications, the homogenisation "power" of ultrasound has been used for achieving uniform and efficient mixing of the chemicals and has been found to be more efficient than using magnetic stirrers or high-shear mixers. In specific operation conditions, ultrasonic irradiation can also result in water sonolysis, which produces radical species (see Chap. 1).

© The Author(s), under exclusive licence to Springer Nature Switzerland AG 2019
B. G. Pollet and M. Ashokkumar, *Introduction to Ultrasound, Sonochemistry and Sonoelectrochemistry*, Ultrasound and Sonochemistry
https://doi.org/10.1007/978-3-030-25862-7_2

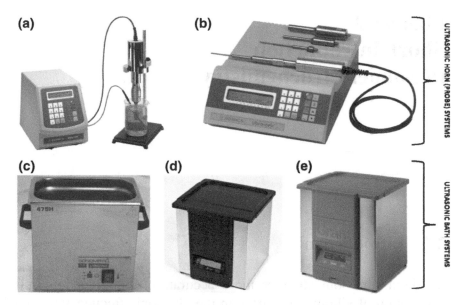

Fig. 2.1 Commercial ultrasonic systems: **a** Vibra-Cell VCX750 ultrasonic probe systems (Sonics and Materials Inc., $f = 20$ kHz, net power output = 50 W); **b** Vibra-Cell VCX130 ultrasonic system with several detachable microtips (Sonics and Materials Inc., $f = 20$ kHz, net power output = 130 W); **c** QS12 ultrasonic bath (Ultrawave, $f = 32–38$ kHz, ultrasonic power = 200 W; heating power = 300 W); **d** 475H Sonomatic ultrasonic bath (Langford Ultrasonics, $f = 40$ kHz, high-frequency peak power = 300 W; heating power = 200 W; **e** XUB18 ultrasonic bath (Grant Instruments, $f = 32–38$ kHz, ultrasonic power = 300 W, heating power = 450 W). Modified from Pollet [2]

Sonoelectrochemistry is the use and study of the effects of power ultrasound in electrochemistry [1]. This concept was first introduced by Moriguchi in the 1930s [3]. Since then, it has been observed in several investigations that ultrasonic irradiation affects not only heterogeneous systems involving the electrode and electrolyte but also homogenous systems that occur in the electrolyte bulk solution [4, 5]. When the electroanalytic is ultrasonically irradiated, it experiences "extreme" conditions caused by acoustic streaming and acoustic cavitation which may give rise to improved and even to new electrochemical reaction mechanisms.

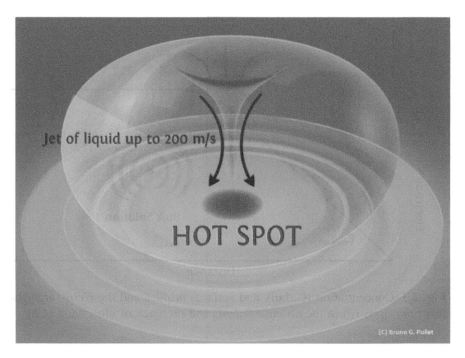

Fig. 2.2 Representation of a cavitation bubble imploding near the electrode surface causing the formation of a high-velocity microjet of liquid hitting the surface. Modified from Pollet [2]

When acoustic cavitation takes place near the electrode surface, the collapse and implosion of the cavitation bubbles leads to the formation of jet of liquids (microjets) directed towards the electrode surface (Fig. 2.2). These microjets can reach speeds up to 200 m s^{-1} and produce local "hot spots" on the electrode surface [6]. All these phenomena contribute in a substantial thinning of the *Nernst diffusion layer thickness* (δ) (Fig. 2.3), an improvement of the overall mass transfer, an increase in electrochemical reaction rates, a cleaning and degassing of the electrolyte and the electrode surface. In most cases, the enhancement in mass transport rate (m_o, see later) is mainly attributed to the turbulent flow created by acoustic streaming and cavitation bubble implosion at the electrode surface.

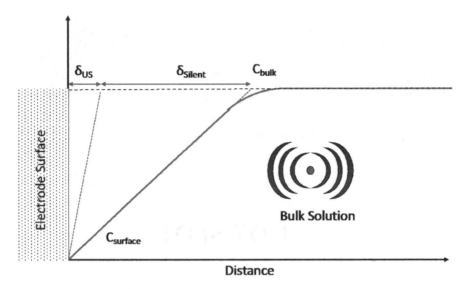

Fig. 2.3 Concentration (C, bulk and surface) profiles and the *Nernst diffusion layer thickness* (δ) in the absence (*silent*) and presence of ultrasound (*US*)

This chapter gives a very brief overview of *sonoelectrochemistry*. For further in-depth information and discussions on *sonoelectro-chemistry*, the reader is invited to read a manuscript entitled "*Power ultrasound in electrochemistry: from versatile laboratory tool to engineering solution*" [1]. The book solely focusses on the use and the effects of power ultrasound in electrodeposition, electroplating, electrocoagulation, electropolymerisation, electroanalysis, corrosion electrochemistry, environmental electrochemistry, organic elec-trosynthesis and electrochemical production of energy nanomaterials.

2.2 The Use of Power Ultrasound in Electrochemistry

As stated previously, the study of the effects of power ultrasound on electrochemical systems is not new as it was first investigated in the 1930s by Moriguchi [3], followed by Schmid and Ehret [7, 8]. This powerful combination was then studied by Yeager and Hovorka [9], Kolb and Nyborg [10], Penn et al. [11] and Bard [12]. In the time of writing and according to *Google Scholar*, over 4500 publications have been published on the subject [13] with the majority of the scientific papers published after a review entitled "Sonoelectrochemistry" by Mason et al. from the Coventry University Sonochemistry Centre of Excellence (UK) in 1990 [14]. In this review paper, the authors highlighted the formidable effects of ultrasonic irradiation on electrochemistry. Since then, there have been many works on the topic as well as a series of numerous reviews. For example, extensive reviews have been compiled by Yegnaraman and Bharathi [15], Walton and Phull [16], Compton et al. [17], Pollet and Phull [18], Walton [19], Brett [20], Klima [21] and Gonzales-Garcia [22], all clearly and systematically describing the extraordinary effects of ultrasonication on electrochemical parameters, redox couples and systems; Walton and Mason [23] and Cognet et al. [24] described the use of ultrasound in organic electrosynthesis; Wadhawan et al. [25] and, Banks and Compton [26] highlighted the use of ultrasound as a powerful tool in electroanalysis, Sáez and Mason [27] focussed on the sonoelectrochemical production of nanomaterials; and Pollet [28] and Islam et al. [29] detailed the sonoelectrochemical and sonochemical production of hydrogen and fuel cell materials. All these reviews showed that combining power ultrasound with electrochemistry led to electroactive species mass transport improvement, efficient electrode surface cleaning, electrolyte sonolysis, improved electrochemical rates and yields, the production of "exotic" (nano)materials, lower overpotentials, enhanced water and soil quality (depollution), better electroanalytical chemical detection limits, and better overall electrochemical process efficiencies.

The first "modern" *sonoelectrochemists* to investigate the use and effects of power ultrasound on electrochemical systems was in the 1990s with Walton et al. [30], Cataldo [31], Yegnaraman and Bharathi [15], Reisse et al. [32], Hagan and Coury [33], Compton et al. [34], Klima et al. [35], Lorimer, Pollet et al. [36], with the Oxford University Electrochemistry Group (UK) who published just over 100 papers in the area.

2.3 Advantages of Sonoelectrochemistry

The use of ultrasonic irradiation in electrochemistry can impart some specific benefits such as: efficient solution and electrode degassing, electrode surface cleaning, and great enhancement of mass transfer of electroactive species from the bulk solution to the electrode surface through the double layer (Fig. 2.3) [1, 4]. These effects are mainly caused by acoustic streaming and acoustic cavitation, i.e. the generation of local steady flows, growth and collapse of microbubbles in the electrolyte. *Sonoelectrochemistry* offers other advantages, for example (i) enhanced electrochemical diffusion processes, (ii) increase in electrochemical rates and yields, (iii) increase in electrode and current efficiencies, (iv) decrease in electrode (over)potentials (E, η), and overall cell voltages (V_{cell}), (v) suppression of electrode fouling and degassing at the electrode surface, (vi) improved electroplated and electrodeposited materials (polymers and metals) in terms of quality, hardness, porosity and thickness and (vii) improved electrode surface activation [1, 4].

Most of these observations are principally attributed to: (a) electrode surface cleanliness, (b) metal depassivation and gas bubble removal at the electrode surface, induced by acoustic cavitation and acoustic streaming, and (c) enhanced mass transfer of electroactive species to the electrode surface yielding a thinning of the diffusion layer thickness (δ) [1, 4]. Figure 2.4 gives a summary of the advantages of using power ultrasound in electrochemistry.

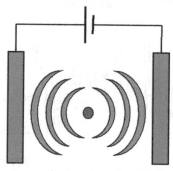

BENEFITS OF SONOELECTROCHEMISTRY

✓ **Enhanced electrochemical diffusion processes**
✓ **Improved electrode kinetics**
✓ **Improved electrode surface activation**
✓ **Better electrode surface cleanliness**
✓ **Lower electrode overpotentials**
✓ **Suppressed electrode fouling**
✓ **Efficient degassing at the electrode surface**
✓ **Enhanced electrochemical rates and yields**
✓ **Improved electrodeposit quality and properties**
✓ **Better electrochemical process efficiencies**

Sonoelectrochemical Cell

Fig. 2.4 Summary of the benefits of power ultrasound in electrochemistry. Modified from Pollet [5]

2.4 Ultrasonic Effects on Electrochemical Parameters

As discussed previously, most of the effects of power ultrasound in electrochemical processes are due to the effects of acoustic cavitation together with microstreaming. There are many studies which conclusively showed that power ultrasound greatly improves mass and heat transport phenomena, although there is little information on the effect of ultrasonication on electrode surface adsorption and electron-transfer processes. Nonetheless, *sonoelectrochemistry* has the ability to promote heterogeneous electrochemical reactions mainly through "extreme" increase in mass transport, interfacial cleaning and thermal effects.

It is now well-accepted in the area that the main effects of power ultrasound in electrochemistry are due to the large increase in mass and heat transfer induced by vigorous and violent solution "mixing" yielding an increased thinning of the diffusion layer thickness (δ) at the electrode surface (Fig. 2.2), a decrease in anodic (E_a) and cathodic (E_c) potentials and overpotentials ($\eta_{a,c}$) and an overall cell voltage

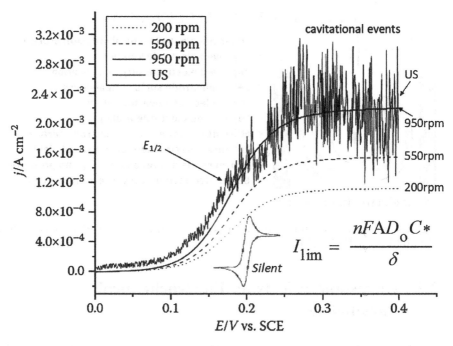

Fig. 2.5 A "sonovoltammogram" (US) and rotating disc electrode (RDE) volta mmograms of a typical quasi-reversible system ($Fe(CN)_6^{3-}/Fe(CN)_6^{4-}$ redox couple) in the potential range [0; +400 mV vs. SCE] at 298 K on a platinum electrode and at several rotation speeds (*silent*: 200, 550 and 950 rpm) and at 20 kHz, recorded at a scan rate of 2 mV/s. Modified from Pollet [38]

(V_{cell}). These observations are mainly due to the continuous surface cleaning and degassing, and surface modification due to erosion, pitting and ablation caused by the implosion of high-energy cavitation bubbles and by acoustic streaming together with microstreaming at or near the electrode surface.

In *sonoelectrochemistry*, several electrochemists have observed sigmoidal "*S*"-shaped electrochemical responses. They named these new types of responses—"sonovoltammograms"—when they studied for quasi-reversible redox systems, e.g. using ferrocene/ferrocenium (Fc/Fc+), hexachloroiridate (III/IV) and hexaammineruthenium

(III/II) chloride, in aqueous solutions (Fig. 2.5, $Fe(CN)_6^{3-}Fe(CN)_6^{4-}$) [1, 4, 6, 14, 16–18, 33–39]. The figure shows several linear sweep voltammograms (LSV) generated by ultrasound (US) and rotating disc electrode (RDE) voltammograms of the well-studied $Fe(CN)_6^{3-}$/ $Fe(CN)_6^{4-}$ quasi-reversible redox couple at 298 K on a platinum working electrode and at several rotation speeds (*silent*: 200, 550 and 950 rpm) and at 20 kHz [6, 36]. They found that the electrochemical responses under ultrasonication are very similar to those obtained by hydrodynamics methods (e.g. polarography, dropping mercury electrode, RDE and wall-jet electrode) [30, 33]. They also concluded that Eqs. (2.1) and (2.2), which relate the limiting current (I_{lim}) or limiting current density (j_{lim}) to the diffusion layer thickness (δ), are valid for "sonovoltammograms" [34, 37].

$$I_{lim} = nFAD_0C^*/\delta \qquad (2.1)$$

$$j_{lim} = nFD_0C^*/\delta \qquad (2.2)$$

where I_{lim} is the limiting current (A), j_{lim} is the limiting current density (A m^{-2}), n is the number of electrons transferred during the electrochemical process, F is the Faraday constant (C mol^{-1}), A is the electrode area (m^2), D_o is the diffusion coefficient of the electroactive species (m^2 s^{-1}), C^* is the bulk concentration of the electroactive species (mol m^{-3}) and δ is the diffusion layer thickness (m) [34, 37].

   Other workers also demonstrated that under ultrasonication, millielectrodes of radii > 1 mm behaved as microelectrodes of radii < 1 mm) [34, 37]. These observations were based on studying and comparing limiting currents generated with a microelectrode under *silent* conditions, comparable to those obtained with a milli-electrode under ultrasonication (20 kHz). Figure 2.6 also shows a series of current spikes on the sonovoltammogram (mostly in the

Levich Equation	*Levich-like* **Equation**
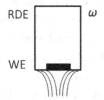	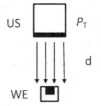
$I_{lim} = 0.620nFAD_o^{2/3}\upsilon^{-1/6}C^*\omega^{1/2}$	$I_{lim} = 0.840nFAD_o^{2/3}\upsilon^{-1/6}r_e^{-1/2}d^{-1/2}A_{uht}^{-1/2}C^*P_T^{1/2}$
I_{lim}: limiting current n: number of electrons F: Faraday constant A: electrode area D_o: diffusion coefficient υ: kinematic viscosity C^*: bulk concentration ω: rotation speed	r_e: working electrode radius A_{uht}: ultrasonic horn tip area d: ultrasonic horn-working electrode distance P_T: ultrasonic (acoustic) power

Fig. 2.6 *Levich equation* for the rotating disc electrode (RDE) and *Levich-like* equation for an ultrasonic (US, 20 kHz) probe facing a working electrode (WE) —"face-on geometry"

plateau region), which are, according to Birkin et al. [39], due cavitational events occurring at the electrode, i.e. jets of electrolyte hitting the electrode surface.

As explained in Chap. 1, homogeneous chemical reactions have been shown and reported to be affected by the sonolysis products mainly by the formation of radical species H• and OH•. Although there is little information available in the literature [40, 41], in a review published by Luche in the 1990s discussed that homogeneous chemical systems should not, in theory, be affected by power ultrasound [41].

In the case of *sonoelectrochemistry*, fundamental investigations on the effect of ultrasonication upon electrode kinetics are scarce. For example, to this date, there have been no systematic studies to elucidate the decrease in cell voltages, anodic and cathodic (over) potentials upon ultrasonication for both quasi-reversible and irreversible redox systems. However, many electrochemists have observed a shift of half-wave potential ($E_{1/2}$) under ultrasonication at several ultrasonic frequencies in the range of [20–800 kHz] [34, 37]. They attributed the shift in $E_{1/2}$ to either (i) hydroxyl radicals (·OH), due to chemical irreversibility in an electrochemical redox system which was quasi-reversible under *silent* conditions, or (ii) a continuous electrode surface cleaning caused by ultrasonication.

In the case of the effect of ultrasonication on the heterogeneous rate constant (k_o), the observations among researchers are somewhat contradictory. For example, Zhang and Coury [42] and Madigan et al. [43] observed that the increase in k_o increases was due to local temperature increases at the electrode surface. However, some electrochemists found that there were no obvious effects of ultrasound on the heterogeneous rate constant [44, 45], and others reported that ultrasound slowed the rate of electron transfer for quasi-reversible redox systems due to electrode surface cleaning [46].

2.5   The *Levich-like* Equation

A study by Pollet et al. [6] showed that, for a "face-on geometry" (i.e. the ultrasonic horn facing the working electrode surface) under ultrasonic conditions, a *Levich-like equation* can be generated (Fig. 2.6) where the limiting current (I_{lim}) is inversely proportional to the ultrasonic horn-working electrode distance (d), the working electrode geometry (r_e) and proportional to the ultrasonic intensity (Ψ) [i.e. the ultrasonic power (P_T) over the ultrasonic horn tip area (A_{uht}), at constant temperature, concentration] according to Eqs. (2.3) and (2.4) [6]:

$$I_{\text{lim}} = 0.84nFAD_{\text{o}}^{2/3}v^{-1/6}r_{\text{e}}^{-1/2}d^{-1/2}C^*\Psi^{1/2} \tag{2.3}$$

$$I_{\text{lim}} = 0.840nFAD_{\text{o}}^{2/3}v^{-1/6}r_{\text{e}}^{-1/2}d^{-1/2}C^*A_{\text{uht}}^{-1/2}P_{\text{T}}^{1/2} \tag{2.4}$$

where I_{lim} is the limiting current (A), n is the number of electrons transferred during the electrochemical process, F is the Faraday constant (C mol^{-1}), A is the electrode area (m^2), D_{o} is the diffusion coefficient of the electroactive species (m^2 s^{-1}), v is the kinematic viscosity (m^2 s^{-1}), d is the ultrasonic horn-working electrode distance (m), r_{e} is the electrode radius (m), A_{uht} is the ultrasonic horn tip area (m^2), C^* is the bulk concentration of the electroactive species (mol m^{-3}), P_{T} is the acoustic power (W) and Ψ is the ultrasonic intensity (W m^{-2}).

Pollet et al. [6] showed that the ultrasonic horn-working electrode positioning is an important factor in determining hydrodynamics parameters, e.g. the limiting current (I_{lim}), the mass transport coefficient (m_{o}), the limiting solution velocity (U_{lim}), the Sherwood (Sh) and the Reynolds (Re) numbers.

2.6   The *Pollet–Hihn* Equation

A few sonoelectrochemists have investigated the acoustic power distribution in various electrochemical reactors operating in the ultrasonic frequency range of (20–100 kHz). There are several methods for determining the ultrasonic or acoustic power, i.e. the power transmitted in the sonoelectrochemical reactor (e.g. the *Besançon* sonoelectrochemical reactor, Fig. 2.7). For example, the aluminium foil, sonoluminescence, calorimetric, chemical dosimetry and laser-sheet visualisation methods can be used to determine the acoustic power.

Another method is to employ electrochemistry as a tool to determine the acoustic power at precise locations in a sono

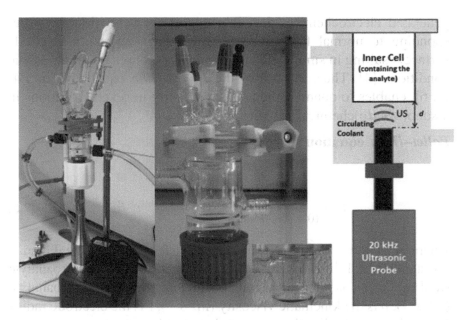

Inner Cell
(containing the
analyte)

US d
Circulating
Coolant

20 kHz
Ultrasonic
Probe

Fig. 2.7 "Face-on geometry" sonoelectrochemical set-up and the *Besançon* sonoelectrochemical cell

(electro)chemical reactor. This new electrochemical method consists of generating limiting currents (I_{lim}) or limiting current densities (J_{lim}) from linear sweep voltammetry (LSV) for quasi-reversible redox couples at low concentrations. By carefully moving the working electrode in the ultrasonic field within the sono(electro)chemical reactor, it is then possible to map out the ultrasonic or acoustic activities within the sono(electro)chemical reactor, especially at areas close to the ultrasonic transducer where intense acoustic cavitation activity exists. To enable comparisons between sonoelectrochemical experiments performed using several ultrasonic and electrochemical equipment, Pollet and Hihn's teams [6] proposed to convert the generated electrochemical data into equivalent velocities, U, corresponding to normal flows directed towards the working electrode surface resulting in similar electrochemical responses under ultrasonic conditions [4]. Thus, the use of the *Pollet–Hihn* equation (Eq. 2.4) [5, 6], enables to quantify the acoustic activity in several locations in a sono(electro)chemical reactor.

***Pollet–Hihn* equation**

$$U = \frac{1}{(0.45 \cdot n \cdot F \cdot C^*)^2} \cdot D^{-4/3} \cdot v^{1/3} \cdot r \cdot j_{\text{lim}}^2 \qquad (2.5)$$

where n is the number of electrons transferred, F is the Faraday constant (C mol^{-1}), C^* is the bulk concentration of the electroanalyte (mol m^{-3}), D is the diffusion coefficient of the electroanalyte (m^2 s^{-1}), v is the kinematic viscosity (m^2 s^{-1}), r is the electrode radius (m) and j_{lim} is the limiting current density (A m^{-2}).

2.7   Applied Sonoelectrochemistry

Since the 1990s, *sonoelectrochemistry* has covered several areas of research from mass transfer measurements in aqueous and non-conventional solvents [47, 48], metallic coating improvements and co-deposition of particles (electrocoagulation, electrodeposition, electroplating, electroless plating and electrocrystallisation) [49–51], accelerated test and metal dissolution (metallic corrosion), metallic coating on non-conductive substrates organic syntheses [52–54],

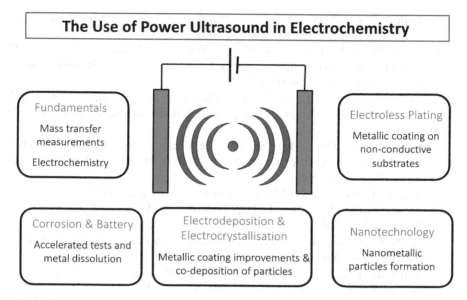

Fig. 2.8 Use of power ultrasound in electrochemistry

polymeric and metallic (nano-)materials syntheses [55, 56], environmental treatments [1], water disinfection [1], analytical sensors [1], to the production of useful gases, e.g. hydrogen [29], and electrolyser and fuel cell electrodes [28, 57, 58] (Fig. 2.8).

References

1. Pollet BG (ed) (2012) Power ultrasound in electrochemistry: from versatile laboratory tool to engineering solution. Wiley. ISBN 978-0-470-97424-7
2. Pollet BG (2014) Let's not ignore the effect of ultrasound on the preparation of fuel cell materials. Electrocatalysis 5(4):330
3. Moriguchi N (1934) The influence of supersonic waves on chemical phenomena III: the influence on the concentration polarisation. J Chem Soc Jpn 55:349
4. Pollet BG, Hihn J-Y (2011) Sonoelectrochemistry: from theory to applications. In: Chen D, Sharma SK, Mudhoo A (eds) Handbook on applications of ultrasound: sonochemistry for sustainability. CRC Press

5. Pollet BG (2008) A short introduction to sonoelectrochemistry. Electrochem Soc Interface Fall 3(27):41
6. Pollet BG, Hihn J-Y, Doche M-L, Lorimer JP, Mandroyan A, Mason TJ (2007) Transport limited currents close to an ultrasonic horn. J Electrochem Soc 154(10):E131
7. Schmid G, Ehret L (1937) Beeinflussung der Metallpassivität durch Ultraschall. Berichte der Bunsengesellschaft für physikalische Chemie 43 (6):408
8. Schmid G, Ehret L (1937) Beeinflussung der Elektrolytischen Abscheidungspotentiale von Gasen durch Ultraschall. Ber Bunsenges Phys Chem 43(8):597
9. Yeager E, Hovorka F (1953) Ultrasonic waves and electrochemistry I: a survey of the electrochemical applications of ultrasonic waves. J Acoust Soc Am 25(3):443
10. Kolb J, Nyborg W (1956) Small-scale acoustic streaming in liquids. J Acoust Soc Am 28:1237
11. Penn R, Yeager E, Hovorka F (1959) Effect of ultrasonic waves on concentration gradients. J Acoust Soc Am 31:1372
12. Bard A (1963) High speed controlled potential coulometry. Anal Chem 35:1125
13. Google Scholar search performed on 08 June 2019—keyword: Sonoelectrochemistry
14. Mason TJ, Lorimer JP, Walton DJ (1990) Sonoelectrochemistry. Ultrasonics 28:333
15. Yegnaraman V, Bharathi S (1992) Sonoelectrochemistry—an emerging area. Bull Electrochem 8:84
16. Walton DJ, Phull SS (1996) Sonoelectrochemistry. Adv Sonochem 4:205
17. Compton RG, Eklund JC, Marken F (1997) Sonoelectrochemical processes: a review. Electroanalysis 9:509
18. Pollet BG, Phull SS (2001) Sonoelectrochemistry—theory principles, and applications. Recent Res Dev Electrochem 4:55–78
19. Walton DJ (2002) Sonoelectrochemistry—the application of ultrasound to electrochemical systems. Arkovic 3:198–218
20. Brett C (2008) Sonoelectrochemistry. In: Arnau Vives A (ed) Piezoelectric transducer and applications, Chap. 15, pp 399–411. Springer, Berlin
21. Klima J (2010) Application of ultrasound in electrochemistry. An overview of mechanisms and design of experimental arrangement. Ultrasonics. https://doi.org/10.1016/j.ultras.2010.08.004
22. González-García J, Esclapez MD, Bonete P, Hernández YV, Garretón LG, Sáez V (2010) Current topics on sonoelectrochemistry. Ultrasonics 50:318–322

23. Walton DJ, Mason TJ (1998) Organic sonoelectrochemistry. Synth Org Sonochem 4:263–300
24. Cognet P, Wilhem A-M, Delmas H, Lyazidi HA, Fabre P-L (2000) Ultrasound in organic electrosynthesis. Ultrason Sonochem 7:163–167
25. Wadhawan JD, Marken F, Compton RG (2001) Biphasic sonoelectrochemistry: a review. Pure Appl Chem 73:1947–1955
26. Banks CE, Compton RG (2003) Ultrasonically enhanced voltammetric analysis and applications: an overview. Electroanalysis 15:329–346
27. Sáez V, Mason TJ (2009) Sonoelectrochemical synthesis of nanoparticles. Molecules 14:4284–4299
28. Pollet BG (2010) The use of ultrasound for the fabrication of fuel cell materials. Int J Hydrogen Energy 35(21):11986
29. Islam MdH, Burheim OS, Pollet BG (2019) Sonochemical and sonochemical production of hydrogen. Ultras Sonochem 51:533
30. Walton DJ, Chyla A, Lorimer JP, Mason TJ (1990) Sonochemical enhancement of phenylacetate electrooxidation. Syn Commun 20:1843
31. Cataldo F (1992) Effects of ultrasound on the yield of hydrogen and chlorine during electrolysis of aqueous solutions of NaCl or HCl. J Electroanal Chem 332(1–2):325
32. Reisse J, Francois H, Vandercammen J, Fabre O, Kirsh-De Mesmaeker A, Maershalk C, Delplancke JL (1994) Sonoelectrochemistry in aqueous electrolyte: a new type of sonoelectroreactor. Electrochim Acta 39:37
33. Hagan RSC, Coury LA (1994) Comparison of hydrodynamic voltammetry implemented by sonication to a rotating disk electrode. Anal Chem 66:399
34. Compton RG, Eklund JC, Page SD, Sanders GHW, Booth J (1994) Voltammetry in the presence of ultrasound. Sonovoltammetry and surface effects. J Phys Chem 98:12410
35. Klima J, Bernard C, Degrand C (1994) Sonoelectrochemistry: effects of ultrasound on voltammetric measurements at a solid electrode. J Electroanal Chem 367:297
36. Lorimer JP, Pollet B, Phull SS, Mason TJ, Walton DJ, Geissler U (1996) The effect of ultrasonic frequency and intensity upon limiting currents at rotating disc and stationary electrodes. Electrochim Acta 41:2737
37. Compton RG, Hardcastle JL, del Campo J (2003) Sonoelectrochemistry, physical aspects. In: Bard-Stratmann (ed) Encyclopedia of electrochemistry; In: Unwin P (ed) Instrumentation and electrochemical chemistry, vol 3, pp 312–327
38. Pollet BG (2019) Does power ultrasound affect heterogeneous electron transfer kinetics? Ultras Sonochem 52:6
39. Birkin PR, Silva-Martinez S (1996) A study of the effect of ultrasound on mass transport to a microelectrode. J Electroanal Chem 416:127

40. Colarusso P, Serpone N (1996) Sonochemistry II.—effects of ultrasounds on homogeneous chemical reactions and in environmental detoxification. Res Chem Intermed 22(1):61
41. Luche J-L (1994) The effect of ultrasound on heterogeneous systems. Ultrason Sonochem 1(2):S111
42. Zhang H, Coury LA Jr (1993) Effects of high-intensity ultrasound on glassy carbon electrodes. Anal Chem 65:1552
43. Madigan NA, Hagan CRS, Zhang H, Coury LA (1996) Sonochemical stripping voltammetry. Ultrason Sonochem 6:S239
44. Marken F, Eklund JC, Compton RG (1995) Voltammetry in the presence of ultrasound: can ultrasound modify heterogeneous electron transfer kinetics? J Electroanal Chem 395(1–2):335
45. Birkin PR, Silva-Martinez S (1997) Determination of heterogeneous electron transfer kinetics in the presence of ultrasound at microelectrodes employing sampled voltammetry. Anal Chem 69:2055
46. Huck H (1987) Die messung der ultraschall-diffusion an einer elektrode und ihre praktische anwendung. Ber Bunsenges Phys Chem 91:648
47. Costa C, Hihn J-Y, Rebetez M, Doche M-L, Bisel I, Moisy P (2008) Transport-limited current and microsonoreactor characterization at 3 low frequencies in the presence of water, acetonitrile and imidazolium-based ionic liquids. Phys Chem Chem Phys 10:2149
48. Pollet BG, Hihn J-Y, Mason TJ (2008) Sono-electrodeposition (20 and 850 kHz) of copper in aqueous and deep eutectic solvents. Electrochim Acta 53:4248
49. Walker R (1993) The effect of ultrasound on electrodeposition and electroplating. In: Mason TJ (ed) Advances in sonochemistry, vol 3. JAI Press, Cirencester
50. Prasad R, Vasudevan P, Seshadri SK (1993) Trans Indian Metall 46(4):247
51. Prasad R, Vasudevan P, Seshadri SK (1994) Indian J Eng Mater Sci 1 (3):178
52. Eberson L, Utley JHP (1982) In: Baizer MM, Lund H (eds) Organic electrochemistry, 2nd edn. Marcel Dekker, New York
53. Torii S (1985) Electroorganic syntheses, methods and applications; Part I: oxidations, vol 15 aus der Reihe: Monographs in modern chemistry, Kodansha Ltd., Tokyo und VCh, Weinheim 1985. 338 Seiten, Preis: DM 138
54. Vassiliev YB, Grinberg VA (1991) Adsorption kinetics of electrode processes and the mechanism of Kolbe electrosynthesis: Part I. Adsorption of carboxylic acids and the nature of the particles chemisorbed on platinum electrodes. J Electroanal Chem 283:359–378
55. Et Taouil A, Lallemand F, Hihn J-Y, Melot JM, Blondeau-Patissier V, Lakard B (2011) Doping properties of PEDOT films electrosynthesized under high frequency ultrasound irradiation. Ultrason Sonochem 18(1): 140–148

56. Gedanken A (2004) Using sonochemistry for the fabrication of nanomaterials. Ultrason Sonochem 11:47–55
57. Rashwan SS, Dincer I, Mohany A, Pollet BG (2019) The Sono-Hydro-Gen process (ultrasound induced hydrogen production): challenges and opportunities. Int J Hydrogen Energy 44(29):14500
58. Pollet BG (2019) The use of power ultrasound for the production of PEMFC and PEMWE catalysts and low-Pt loading and high-performing electrodes. Catalysts 9(3):246